FARIDA ALLAL
FATIMA ZOHRA HELIFA
KHADIDJA LAMRI

Metronidazole co-crystals

FARIDA ALLAL
FATIMA ZOHRA HELIFA
KHADIDJA LAMRI

Metronidazole co-crystals

ScienciaScripts

Imprint

Cover image: www.ingimage.com

This book is a translation from the original published under ISBN 978-620-6-71730-0.

Publisher:
Sciencia Scripts
is a trademark of
Dodo Books Indian Ocean Ltd. and OmniScriptum S.R.L publishing group

120 High Road, East Finchley, London, N2 9ED, United Kingdom
Str. Armeneasca 28/1, office 1, Chisinau MD-2012, Republic of Moldova, Europe
Managing Directors: Ieva Konstantinova, Victoria Ursu
info@omniscriptum.com

Printed at: see last page
ISBN: 978-620-8-57888-6

INTRODUCTION

Bacterial infections are caused by a variety of micro-organisms and are the cause of some of the most fatal diseases and widespread epidemics. A large number of antibiotics have been developed to treat them, but unfortunately their excessive use has led to the emergence of bacterial multi-resistance, which has become a public health problem, particularly with the increase in infectious diseases that are difficult to treat. At the same time, the discovery and development of new antibiotics has slowed considerably, and the drugs we currently have are becoming increasingly ineffective. This huge gap between therapeutic needs and the actual number of new drugs is of concern to researchers in galenics and has prompted them to find innovative strategies to combat the emergence of resistant bacteria, with aim of therapeutic efficacy by increasing the bioavailability of the active ingredient, minimising the risk of emergence of resistant strains and preventing reinfection (S. Yaghoubi et al,2021).The strategies most commonly used to improve the therapeutic effect and antibacterial activity of active pharmaceutical ingredients involve exploring their solid forms, in particular polymorphs, solvates/hydrates, salts and co-crystals. Among these different approaches considered to remedy the problem of bacterial resistance, the strategy of co-crystallisation of antibiotics with coformers appears to be a promising solution for improving physicochemical properties, inhibiting resistance and enhancing antibacterial activity (K Zheng et al, 2019). It is in this context that this thesis is set, the main objective of which is to contribute to the search for solutions to combat bacterial resistance to antibiotics, through the co-crystallisation of an active principle widely used in the treatment of anaerobic infections and in the treatment of patients suffering from infections caused by Trichomonas. This is metronidazole, an antibiotic belonging to the 5-nitroimidazole family, in cocrystal form, composed of a combination of the antibiotic with a coformer, which improves its physico-chemical properties, increases its solubility and antibacterial effect, and restores bacterial sensitivity (A Thayyil et al, 2020). The choice of coformer is therefore based on its chemical structure, the presence of functional groups, proton donors and acceptors, capable of establishing non-covalent intermolecular interactions, such as hydrogen bonding, Van Der Waals forces, etc. and π-π interactions (M Singh et al,2023). The different coformers chosen are arginine, proline the leucine, the nicotinamide and the paracetamol. In order address all of these objectives, this thesis has been divided into three chapters: The first chapter presents a bibliographical summary, explaining the concept of co-crystallisation, its advantages and some of the work cited in the literature that has led to promising results in the field of improving the physicochemical properties and antibacterial effect of antibiotics. The second chapter describes the equipment used, the experimental protocols for synthesis and characterisation, carried out using various techniques, namely differential scanning calorimetry (DSC), powder x-ray diffraction (PXRD), scanning electron microscopy (SEM) and fourier transform infrared (FTIR).The third chapter deals with the results of the synthesis and evaluation of antibacterial activity in vitro. The most significant results of this work and possible future prospects are summarised in the general conclusion.

CHAPTER I
SUMMARY BIBLIOGRAPHY

I. Introduction

Oral administration of drugs is favoured because of its many advantages, such as ease of use, absence of pain on administration, reduced risk of infection and production costs, which are often lower than for other types of administration.

However, the constraints associated with the physiology of this route, which requires it to pass through several barriers, such as, the oral cavity, stomach and intestine, give rise to various problems during the pharmacokinetic stages (S. Aitipamula et al, 2012). These problems are related to the possibility of enzymatic degradation of the drug, poor solubility, which poses a great problem, regarding its dissolution, absorption and consequently, bioavailability (D. Bučar et al, 2013). Indeed, good solubility in water and human body fluids, is an indispensable condition for an active ingredient to be therapeutically effective. This is particularly crucial for the oral administration of drugs. However, new active ingredient molecules are becoming increasingly complex and therefore less soluble and less bioavailable in the bloodstream. In order to respond to this problem, it is essential for researchers in galenics to evaluate all the possibilities for improving solubility and developing appropriate formulations, by developing new strategies (**figure I.1**). The 1st strategy is based on physical modifications of the crystalline state and the reduction of crystal size (S. Aitipamula et al, 2012).The 2nd approach is based on chemical modifications such as the formation of more water-soluble salts from a poorly water-soluble active ingredient, the molecular dispersion of the active ingredient in a polymer matrix, the development of nanoparticles, micro- and nano-emulsions, the formation of solvates and the formation of co-crystals. It is on this last approach the present work focuses (J. Bauer et al, 2001).

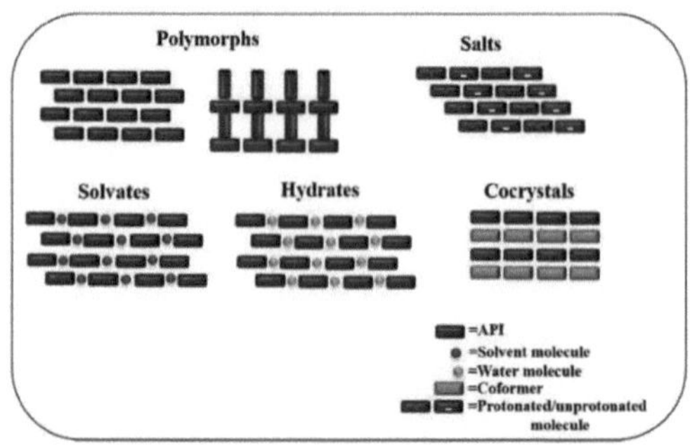

Figure I.1. Different solid forms of active pharmaceutical ingredients (E. Korotkovaa, et al 2014).

I .1. Cocrystallisation

In the pharmaceutical industry, the co-crystallisation of an active ingredient is an alternative to the formation of salts, as it is likely to involve a larger number of molecules than salts. Co-crystallisation is a useful and original process for improving the bioavailability and dissolution speed of drugs. It is the PA/coformer interaction, leading to a crystalline structure consisting of at least two constituents (D. Bučar et al, 2013). It results in a modification of the physicochemical properties of the active ingredients, leading to a drug with intrinsically stable molecular activity (R. Banerjee et al, 2005).

I.2. Cocrystals

Co-crystals are single-phase crystalline solids composed of two or more molecular and/or ionic compounds in a stoichiometric ratio, of which they are neither solvates nor salts (R. Banerjee et al, 2005). Their crystalline structure is different from that of the starting molecules. This directly affects the solubility of a given solid in solution, which makes the drug bioavailable in the body and can therefore be administered using conventional techniques (P. Billot et al, 2013). Pharmaceutical co-crystals are therefore composed of an API and a co-crystallising agent (coformer), generally a small organic compound, which may be another drug or a non-toxic molecule (Biscaia et al., 2021). The cocrystals are maintained by supramolecular heterosyntheses, occurring between the functional groups, in this case carboxylic acid, aromatic nitrogen, amide and alcohol, through interactions of non-covalent nature such as hydrogen bonding, Van der Waals forces, lipophilic contacts and π···π interactions (A. Thayyil et al, 2020).

I.3. Methods for preparing co-crystals

SchemeI.1. Methods for the preparation of cocrystals (M. Singh et al, 2023).

I.4. Active principle

An active ingredient is a substance contained in a medicinal product. However, most of today's active ingredients are prepared by chemical synthesis or semi-synthesis from natural substances (J. Dangoumau et al, 2006).

Table I.2: Examples of some medicines and their active ingredients (Original, 2023).

Drug name	Active ingredient(s)
Efferalgan Flagyl Biofenac	Paracetamol Metronidazole Diclofenac sodium

I.5. Coformer

A component which interacts non-ionically with the API in the crystal lattice, which is not a solvent (including water), and which is generally non-volatile. Coformer selection plays an important role in the choice of final cocrystal attributes. Coformers have the ability to modulate stability and solubility of API when prepared as a cocrystal, by inducing changes in its crystal structure. Factors such as the type of functional group, pKa, their physical form and molecular size must be taken into when forming the cocrystal. The usefulness of chemicals as coformers depends on the ability of the molecules to hydrogen bond with the API. According to Etter's rule, hydrogen bonding is formed if good proton donors and acceptors participate in the formation of this bond (M. Singh et al, 2023).

I.6. Bacteria

The bacteria shown in **figure I.2** are unicellular microorganisms. They are able to reproduce autonomously unlike viruses, which need to hijack the machinery of a cell in order to reproduce. Bacteria vary in size from 1 to 10 μm, and weigh from 10-12 grams. They can be found everywhere. As a prokaryote, the structure of the cell is simple. The inner volume, called the cytoplasm, bounded by the plasma membrane. The membrane controls the flow of nutrients into and out of the bacterium and serves as a support for certain enzymes. This volume is continuous and generally contains no complex secondary structures. All chemical reactions that are sources of energy or enable the bacterium to maintain and multiply take place in the cytoplasm. They may, however, be localised, on the membrane for example. Bacteria do not have a nucleus bounded by a membrane to isolate the genetic material. Genetic information is carried by deoxyribonucleic acid (DNA). This DNA is assembled in the form of one or more chromosomes. The structure of these chromosomes is a double helix of circular DNA coiled together by supercoiling. Beyond the cytoplasmic or plasma membrane, bacteria have a wall made up of peptidoglycan. This ensures the mechanical cohesion of the cell. Some bacteria have only a thick, complex wall. This remains permeable to Gram stain and the corresponding bacteria are known as Gram-positive. Other bacteria have a simpler peptidoglycan wall and, above all, a thicker, more complex cell wall. outer membrane. The outer membrane is impermeable to Gram dye, and bacteria with this structure are Gram-negative.

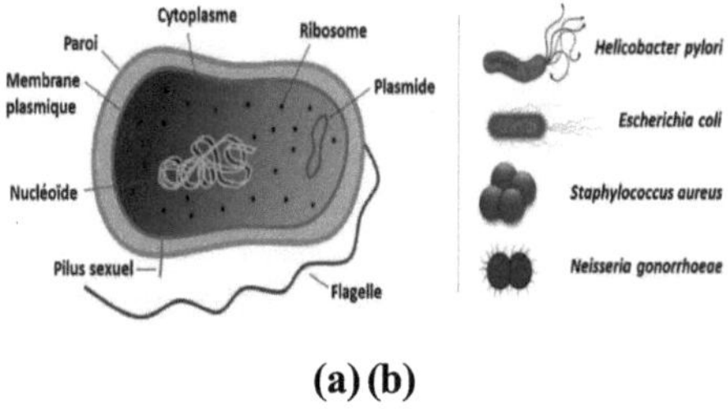

(a) (b)

FigureI.2. Representation of the composition a bacterium (a). Representation of the different morphologies of the best-known bacteria (b) (M.Bousekraoui, 2017).

I.7. Antibiotics

Antibiotics are natural antibacterial agents of biological origin, produced by fungi and/or various bacteria. However, some are produced synthetically, and many are semi-synthetic derivatives prepared by chemical modification of natural products. Some can act on the majority of Gram-positive and Gram-negative pathogenic species, and are known as "broad-spectrum", while others have a more limited action (antibiotics for Gram-positive or Gram-negative bacteria), or even a very narrow spectrum (antistaphylococci, antituberculosis).

Antibiotics have the advantage, when reaching their target, of destroying a structure that is specifically bacterial, by acting at different levels of this structure. Antibiotics with an identical basic chemical structure, which gives them the same antibacterial mechanism of action, are classified in the same family, but they can be differentiated by their spectrum of activity. We distinguish between :

• Antibiotics that inhibit peptidoglycan synthesis: such as β-lactams

• Antibiotics active on membrane envelopes, such as gramicidins.

• Nucleic acid inhibiting antibiotics: such as rifamycins, quinolones and imidazoles.

• Antibiotics that inhibit protein synthesis: such as tetracyclines,

• Antibiotics that inhibit folate synthesis: family of sulphonamides and 2-4-diaminopyrimidines.

In this work, we focused on Metronidazole (MTZ). It is a synthetic antibacterial derived from azomycin, a nitroimidazole produced by the genera Actinobacteria and Proteobacteria. This active ingredient was first used in 1960 to treat trichomoniasis, an infection caused by the protozoan Trichomonas vaginalis. In addition, metronidazole has been successfully used to treat dysentery and excess liver production caused by the intestinal parasite Entamoeba histolytica. It has also been effective against Giardialamblia, another intestinal parasite that causes malabsorption, epigastric pain and certain anaerobic bacterial infections. The main organ responsible for the metabolism of metronidazole is the liver, where it is hydroxylated, acetylated and/or conjugated to glucuronides. Finally, the majority of metabolic waste products are excreted by the kidneys (A. Hernández Ceruelos et al, 2019).

It is important to mention the biopharmaceutical classification system (BCS) when discussing drug solubility and permeability. The CBS divides drugs into four classes based on solubility and permeability properties, including Class I (high solubility and permeability), Class II (low solubility and high permeability), Class III (high solubility and low permeability) and Class IV (low solubility and permeability), which links the physicochemical properties of drugs to oral bioavailability. Low solubility and/or permeability prevents the absorption and bioavailability of these class II, III and IV drugs (Z. Wang et al, 2022). As it is both extremely soluble and permeable, metronidazole is SCB class I (C. Rediguieri et al, 2011).

I.8. Mechanism of action of metronidazole

Metronidazole diffuses throughout the body, inhibiting protein synthesis by interacting with DNA, causing a loss of DNA helical structure and strand breaks. As a result, it causes cell death in susceptible organisms.The mechanism of action of metronidazole involves a four-stage process. The first stage is entry into the body by diffusion through the cell membranes of anaerobic and aerobic pathogens. However, the antimicrobial effects are limited to anaerobes. The second stage involves reductive activation by intracellular transport proteins by modifying the chemical structure of pyruvate-ferredoxin oxidase-canalase. The reduction of metronidazole creates a concentration gradient in the cell, leading to the uptake of more drug and promoting the formation of cytotoxic free radicals. The third stage, interactions with intracellular targets, is achieved by cytotoxic particles interacting with the host cell's DNA, resulting in DNA strand breaks and fatal destabilisation of the DNA helix. The fourth step is the breakdown of the cytotoxic products (N. Kaakoush et al, 2009).

I.9. Antibacterial activity of co-crystals

Although there are few publications on antimicrobial activity involving cocrystals, an in-depth study on the use of cocrystallisation is needed to remedy bacterial resistance by modifying the characteristics of the active ingredient (M.Bashimam, H.El- Zein, 2022).

There are four types of bacterial resistance to antibiotics:

- Enzyme inactivation by the secretion of an enzyme.
- Active efflux.
- Modification of the target.
- Reduced permeability (porins) to the antibiotic.

I. 10 Previous work

A review of the literature enabled us to summarise the little previous work carried out on the co-crystallisation of metronidazole, with a view to increasing its antibacterial effect against various micro-organisms. J. P. F. Whelan et al tested the antibacterial activity of MTZ against Bacteroides fragilis strains isolated from human lesions using the blood agar dilution method. Metronidazole was found to be active against Bacteroides fragilis strains isolated from human lesions. The minimum inhibitory concentration (MIC) and minimum bactericidal concentration (MBC) are equivalent, ranging from 0-16 to 2-5 µg/ml (J. Whelan et al, 1973).Hachem C.Y et al determined the minimum inhibitory concentration (MIC) of ampicillin, clarithromycin and metronidazole against 122 clinical isolates of Helicobacter pylori using microdilution and disc diffusion (E-test). A complete correlation between the results of the two methods was obtained in the presence of ampicillin and clarithromycin. However, tests with metronidazole showed differences in the MICs obtained by the two methods used, with values of 8 and 32µg/ml, obtained by microdilution and diffusion respectively (C. Hachem et al, 1996). A study by J.Li, et al, looked at the possibility of improving the bioavailability of MTZ using a new cocrystal, by combining it with ethyl gallate as a coformer. The results obtained showed that the synthesised cocrystal has a higher solubility than pure PA, leading to faster dissolution and higher bioavailability. In addition, MTZ-EG cocrystal is physically stable in water for at least 24 hours. Its improved solubility and solid state stability during dissolution make it a suitable candidate for further development into an oral solid dosage form with improved biopharmaceutical performance (J. Li et al, 2021).K. Zheng et al (2019) developed, characterised and evaluated in vitro and in vivo crystallisation of acid gallic acid-metronidazole according to the following experimental protocol: To a stirred aqueous solution (20 mL) containing MTZ (0.8558 g, 5 mmol) at 70°C, an aqueous solution (10 mL) of GA-H_2O (0.9407 g, 5 mmol) was added dropwise.

The colour of the solution changed from colourless to yellow with the addition of GA-H_2O. After stirring the reaction for 1 H, the resulting solution was filtered and dried in vacuo.The results obtained showed the effect of co-crystallisation in improving the pharmacokinetic characteristics of drugs. The experimental synthesis study was combined with modelling of drug structure and properties to provide a new methodological approach for the development of new drug cocrystals, contributing to the development of new solid forms and formulations with improved performance (K. Zheng et al, 2019). K. Zheng et al (2020) synthesised a new cocrystal of the antimicrobial drug MTZ in the presence of a natural compound, PYR. The cocrystal formed was characterised by X-ray diffraction, thermal analysis, UV-vis and FTIR. In vitro dissolution results showed that the MTZ-PYR cocrystal has a higher dissolution rate than pure MTZ under acidic and neutral conditions (K. Zheng et al, 2020). Beena et al synthesised metronidazole/triazole cocrystals and evaluated their antibacterial activity. The results showed powerful antibacterial effects at low concentrations against gram-positive and gram-negative bacteria compared with a reference antibacterial (Beena et al, 2009).

CHAPTER II

EXPERIMENTAL

II.1 Introduction

The aim of this work is to synthesise and characterise new antibacterial molecules, based on active drug ingredients and organic molecules, which act as coformers, and to test their ability to combat bacterial strains, particularly those resistant to standard antibiotics. The active ingredient used in this work is metronidazole. It was mixed with a number of coformers chosen for their chemical structure, containing several heteroatoms capable of establishing interactions with the proton donor and acceptor functional groups of metronidazole. A summary diagram of the experimental procedure followed is shown below:

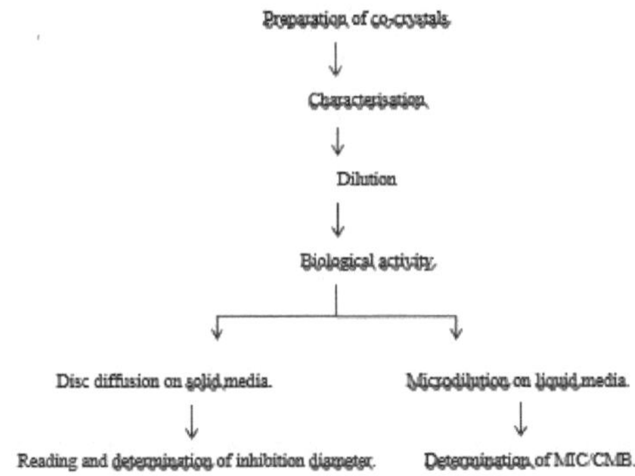

DiagramI.2. Summary diagram of the experimental procedure followed.

II.2 Materials and methods

II.2.1. Chemicals used

The chemicals used, which are listed in tables II.1, have not undergone any subsequent purification.

TableII.1. Chemicals used and their properties.

	Gross formula	Molecular weight (g/mol)	Purity(%)	Source
Metronidazole	C6H9O3N3	171.16	99	Alfa Aesar
Paracetamol	C8H9NO2	151.16	98	Alfa Aesar
Nicotinamide	C6H6NO2	122,12	99	Biochempharma
L-Leucine	C6H13NO2	131.17	98	Sigma-Aldrich
L-Proline	C5H9NO2	115.13	98	Sigma-Aldrich
L-Arginine	C6H14N4O2	174.20	98	Sigma-Aldrich

The physico-chemical properties of the products used in this work are shown in (**tableII.2**).

TableII.2. Physico-chemical properties of the chemicals used.

Structure	Number of hydrogen bond donors	Number of acceptors liaison hydrogen	PKa	Solubility	Physical condition	Melting point
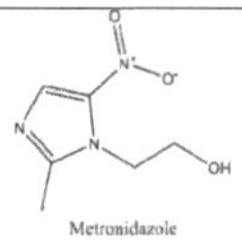 Metronidazole				Very soluble in water,		
	1	4	2.38	soluble in ethanol, ether and chloroform; soluble in dilute acids; sparingly soluble in dimethylformamide	White to pale yellow crystalline powder	160 °C
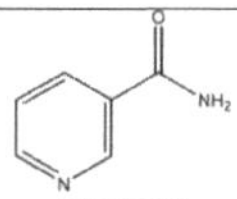 Nicotinamide				Very soluble in water 1 g soluble in 1 mL water		
	1	2	3.35		White powder	128 °C
				Soluble in butanol, chloroform		

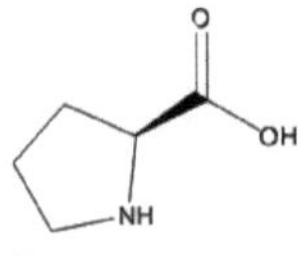 Proline	2	3	1.99/10.60	Very soluble in water Soluble ethanol, acetone, benzene, insoluble in ether, propanol	Dry, white powder	221 °C
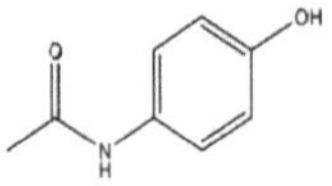 Paracetamol	2	2	9.38	Soluble in water boiling Soluble in alcohol, methanol, ethanol, acetone, ethyl acetate, ether, Insoluble in petroleum ether, pentane, benzene	White powder	169 °C
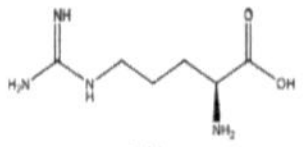 Arginine	4	4	2.24	Soluble in water Insoluble in ethyl ether	White powder	223 - 224 °C
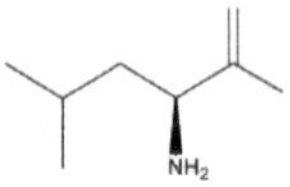 O OH Leucine	2	3	2.35	Soluble in water Soluble in acid Acetic ethanol	White powder	286

II.2.2. Microorganisms tested

The microorganisms chosen for this study, summarised in **Table II.3,** are reference strains.

TableII. 3. General information on the bacterial strains used (M.Bouskraoui et al, 2017).

Strain bacterial tested	Bacteriological character	Habitats	Pathogenicity	Reference
Staphylococcus aureus	Gram+ Optional anaerobic	Man is the main contaminate surfaces, air and water.	Infections Superficial or deep suppurative infections: skin, soft tissue, muscle, bone. Toxi infection: due to the synthesis of different toxins by certain strains.	ATCC 25923
Echerichia coli	Gram - Optional anaerobic	Normal host of the digestive tract.	Enterocolic infections, urinary tract infections, food poisoning, intra-abdominal infections, neonatal infections (meningitis).	ATCC 25922
Pseudomonas aeroginosa	Gram - Anaerobic	bacterium, source of nosocomial infections of exogenous origin and of endogenous origin.	Community infections: eye, ENT, skin. Care-associated infections: pneumonia, urinary tract infections, post-operative infections.	ATCC 27853
Klebsiella pneumoniae	Gram - Optional anaerobic	Natural cavities, in particular the digestive tract and airways superior.	Nosocomial and community-acquired infections: bronchopulmonary and urinary tract. Purulent meningitis and sepsis...	ATCC 70603

II.3. Experimental method

II.3.1. Mechanical synthesis of co-crystals

II.3.1.1. Liquid grinding (LAG)

Liquid-assisted grinding consists of adding a small volume of solvent to the PA/coformer mixture to enable a cocrystal to be formed. The solvent in acts as a catalyst, enabling the establishment of the interactions required for co-crystallisation (A. Thayyil et al 2020). In order to prepare the various cocrystals used in this work, we proceeded as follows:

Using an Ohaus analytical balance accurate to $10^{(-4)}$ g, masses of metronidazole and the various coformers were weighed in a 1:1 stoichiometric ratio (Table II.4). The weighed masses m_1 and m_2 were mixed and ground by wet grinding using a pestle and mortar for 30 min, assisted by addition of 1 ml methanol. Once synthesised, they were placed in an oven to allow the solvent to evaporate.

TableII.4. Masses m_1 and m_2 of metronidazole and the various coformers.

Cocrystals	Cocristal 1 (MTZ / Paracetamol)	Cocristal 2 (MTZ / Nicotinamide)	Cocristal 3 (MTZ / L. Leucine)	Cocristal 4 (MTZ / L. Proline)	Cocristal 5 (MTZ / L. Arginine)
m1/n1 (mg/mmol)	52.5 / 0.3	52.5 / 0.3	73.7 / 0.8	137.3 / 0.8	136.6 / 0.8
m2/n2 (mg/mmol)	45.8 / 0.3	37.7 / 0.3	105.6 / 0.3	92.4 / 0.8	141.8 / 0.8

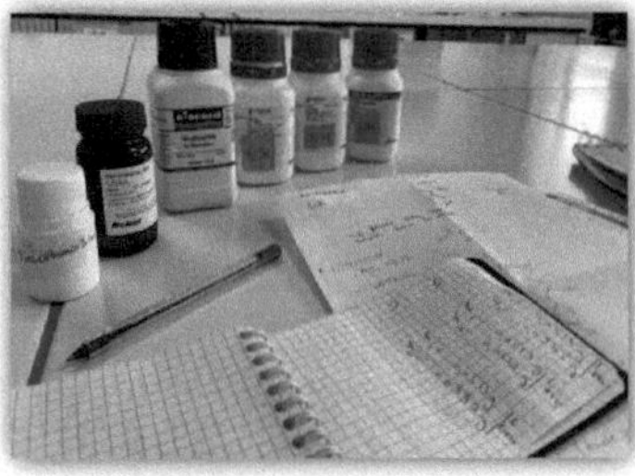

FigureII.1. Calculation of concentrations (Original,. 2023). **FigureII.2.** Liquid assisted milling LAG (Original,. 2023).

II.4. Characterisation of co-crystals

Crystallinity plays an important role in the stability and clinical performance of pharmaceutical cocrystals, and the characterisation of target cocrystals in solids is generally determined by these techniques:

II.4.1 Differential scanning calorimetry DSC

Differential Scanning Calorimetry (DSC) is a method of measuring the temperature of a surface. Calorimetry is a thermal analysis technique used to characterise changes in the state, phase or structure of a material. It is used to measure the heat flux exchanged by a sample and the oven relative to an inert reference as a function of temperature and time, when they are subjected to the same heating or cooling programme (where T varies linearly with t). The temperature difference is converted into a heat flux difference in mW. Using this technique, we can identify the thermal events that a compound may undergo. These events may be reversible, such as melting or recrystallisation, or irreversible, such as degradation or chemical reaction on heating. To carry out the differential scanning calorimetry (DSC), we used a Setaram 131 DSC apparatus (**Figure II. 3**) with aluminium crucibles of 30 µl volume and Platinum thermocouples. The sample size was approximately 1 to 2 mg. The analysis was carried out in a nitrogen atmosphere at a heating rate of 2°C/min, from room temperature (20°C) up to 250°C. Temperatures were calibrated using indium, which has a melting temperature of 156.6°C and an enthalpy of fusion of 28.5 J/g.

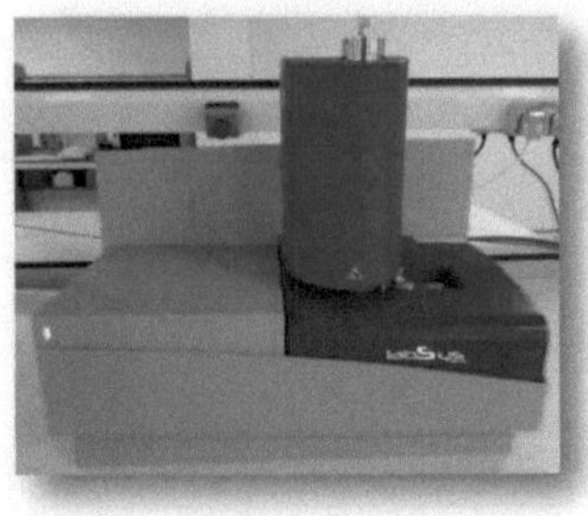

Figure II.3. DSC used to measure heat flux (Original , 2023).

II.4.2. X-ray diffraction

X-ray diffraction is one of the physicochemical analysis used to characterise crystal lattices. X-rays, like all electromagnetic waves, cause a displacement of the electron cloud relative to the nucleus in atoms; these induced oscillations result in the re-emission of electromagnetic waves of the same frequency. Depending on the direction in space, a large or small flow of X-ray photons takes place, and these variations in direction form the phenomenon of X-ray diffraction. Using Bragg's law, we can determine the reticular distance For X-ray diffraction analysis, we used an EMPYREAN diffractometer (**Figure II. 4**). X-rays were emitted using Cu(Kα) radiation at room temperature over a range of 2 theta between 2 and 50°, with a step size of 0.01° and a scanning speed of 2°/min.

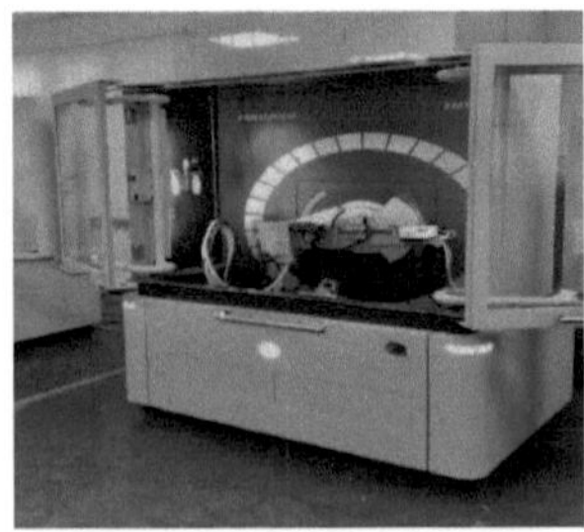

FigureII.4. EMPYREAN type diffractometer (Original, 2023).

II.4.3. Scanning electron microscopy SEM

Scanning electron microscopy is a technique used to visualise the dispersive behaviour of a material. The principle involves scanning the surface of the sample with a very fine beam of electrons that scans the surface point by point. Under the impact of this accelerated electron beam, backscattered electrons and secondary electrons emitted by the sample are selectively collected by detectors that transmit a signal to a cathode-ray screen whose scan is exactly synchronised with the scan of the object. This technique us information about the topography, morphology and composition of the sample. The equipment used is a Thermo Fischer SCIO2S environmental scanning electron microscope. Scanning electron microscopy analysis shows the shape and size of the particles obtained on a microscopic scale.

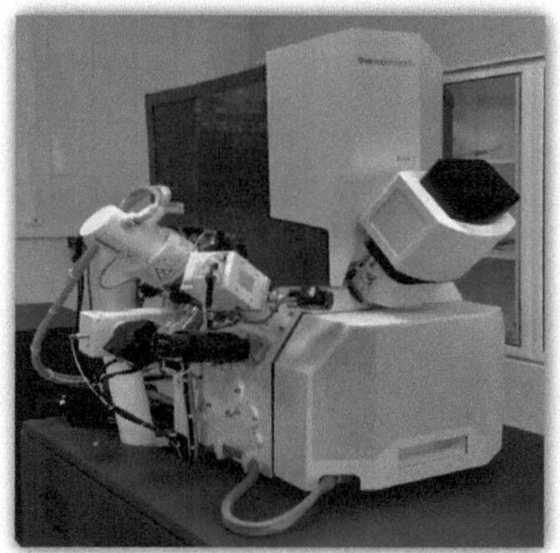

Figure II.5. Thermo Fischer SCIO2S environmental scanning electron microscope (Original, 2023).

II.4.4. Infra-red IR

Infrared spectroscopy is mainly used to study the interactions between molecules, by analysing the profile of the vibratory mode, i.e. the position, width and intensity of the spectral bands. The position of the peak or band not only indicates the presence of a particular group, but also gives a good idea of the environment affecting it. It is well known that intermolecular forces due to hydrogen bonding interactions cause a remarkable change in some of the vibrational modes, making it possible to study the interactions. The formation of hydrogen bonds is of vital importance in many industrial processes, and plays a central role in biological processes at the molecular level. It is responsible for the structural reorganisation of mixed molecules, and also determines the structure and properties of many biological molecules and macromolecules.Infrared spectroscopy plays a crucial role in the study of hydrogen bonding. To confirm the co-crystallisation of our active ingredient, we carried an infra-red analysis using a Jasco FT/IR-4200 spectrophotometer, shown in **Figure II.6**.

FigureII.6. Jasco FT/IR-4200 spectrophotometer (Original, 2023).

II.5. Determination of the antibacterial activity of the cocrystals produced

II.5.1. Dissolution of the pure active ingredient and its co-crystals in DMSO

Using an Ohaus-type analytical balance, we weighed well-defined masses of the pure active ingredient and the cocrystal (**Table II.5).** The cocrystals were then dissolved in 5 ml of DMSO (Dimethylsulfoxide) and homogenised using a Vortex shaker (**Figure 7**).

TableII.5. Masses of pure active principle and cocrystals.

Weighed material	Mass (mg)
Pure metronidazole	61
Cocristal 1	60
Cocristal 2	61.4
Cocristal 3	61.8
Cocristal 4	61.4
Cocristal 5	60.7

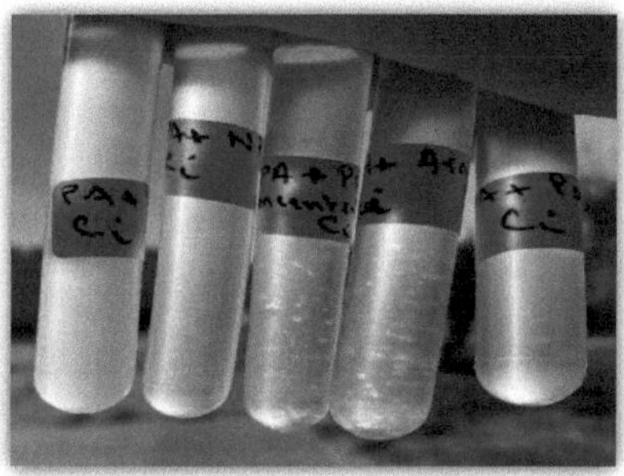

FigureII.7. Dissolution of cocrystals in DMSO (Original, 2023).

II.5.2 Preparation of cocrystals diluted 1/2, 1/5 and 1/10

In order to obtain a concentration range, we diluted the stock solution into three daughter concentrations, i.e. concentrations of 1/2, 1/5 and 1/10 respectively. In addition, the positive control was also diluted. The same volume (50µl) of DMSO (Dimethylsulfoxide) was used as the negative control. The volumes we used are as follows: 1\2 : 25ul crude cocrystal+ 25µl DMSO. 1\5 : 10ul cocrystal crude+ 40µl DMSO. 1\10 : 5ul cocrystal crude+ 45µl DMSO. 50µl crude cocrystal.

II.6. Study antibacterial activity

Antibacterial activity was assessed using two methods:

- The solid-state diffusion method, which consists observing the zone inhibition of bacterial growth.
- The liquid dilution method, which is based on determining the minimum inhibitory concentration (MIC) and minimum bactericidal concentration (MBC) of the co-crystals.

II.6.1. Method of distribution on disc

II.6.1.1. Preparation of the preculture

Antimicrobial tests need to be carried out on young bacterial cultures that are 18 to 24 hours old and in an exponential growth phase. The strains are regenerated by introducing the bacterial species into a culture medium (CM).

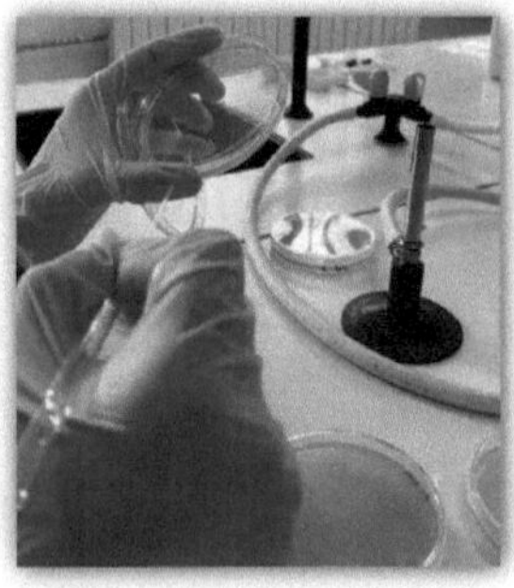

FigureII.8. Replication of bacterial strains (Original, 2023).

II.6.1.2. Preparation of the bacterial suspension

After growing the bacteria in MH agar medium, we selected 3 to 5 well-isolated, identical colonies. These were transferred to 5 ml of sterile physiological water and shaken with a vortex mixer. The bacterial suspension was standardised to a concentration of 10^6 CFU, corresponding to an absorbance of 0.08-0.1 measured using a UV spectrophotometer at a wavelength of 625 nm.

II.6.1.3. Seeding

Prepare the MH in accordance with the manufacturer's instructions and allow it to cool to close to room temperature. Spreading the bacterial suspension: Using a sterile swab, spread the bacterial suspension over the entire surface of the CBM. It is important to ensure even distribution. Incubation: Incubate petri dishes at 37 °C for 24 hours to allow growth. bacterial.

II.6.1.4. Preparing the discs

Once the inoculation of bacteria in the MH culture medium has been established, Wattman No. 3 paper are placed on the agar surface using sterilised forceps, then soaked with 10 µl of crude cocrystals and diluted. Three soaked discs of each concentration were placed in the same petri dish. Petri dishes containing all the concentrations of the different cocrystals synthesised were incubated at 37 C for 24 hours. Once incubation is complete, the inhibition diameters of the different bacteria tested are measured using a calliper.

FigureII.9. Preparation of cocrystal-soaked discs (Original, 2023).

II.6.2. Liquid micro-dilution method

The microdilution method was used to determine the antibacterial efficacy of metronidazole cocrystals in a liquid medium. It was used to assess the minimum bactericidal concentrations (MBC) and minimum inhibitory concentrations (MIC) for the cocrystals used in this work.

II.6.2.1. Determination of the minimum inhibitory concentration (MIC)

The MIC is the lowest concentration of antibiotic that inhibits visible growth after incubation period of 18 to 24 hours. It is used to define the sensitivity or resistance of bacterial strains to antimicrobials (Kalban et al, 2008).

II.6.2.2. Determining the minimum bactericidal concentration (MBC)

The minimum bactericidal concentration (MBC) is the concentration of an antimicrobial agent that leaves no more than 0.01 surviving germs. (Moroh et al, 2008). This is the lowest concentration necessary to kill the bacteria present in a culture.

II.6.2.3. Preparation of microplates

The MIC (Minimum Inhibitory Concentration) of the microorganisms studied was assessed a sterile 96-well microplate, using a single plate for each Cocristal. To do this, proceeded as follows (**Figure II.10**):

The wells were inoculated with 100 µl of liquid Mueller Hinton broth, with the exception of the rows (B, D, F and H), which were left empty to avoid possible contamination. To wells 1 to 6, 100 µl of the different cocrystals synthesised were added. Wells 7 to 12 were kept as a positive control using the pure active ingredient, metronidazole. A serial dilution of the order of 1/2 was carried out on the various filled wells. A volume of 100 µl of the four bacteria tested in this work, E.coli, P.aeroginosa, S.aureus and K.pneumoniae were added to the wells in question. The final volume: broth, cocrystal and bacteria is 200 µl. Once the microplate is prepared, incubation for 18 hours at 37°C is essential for reading the expected results.

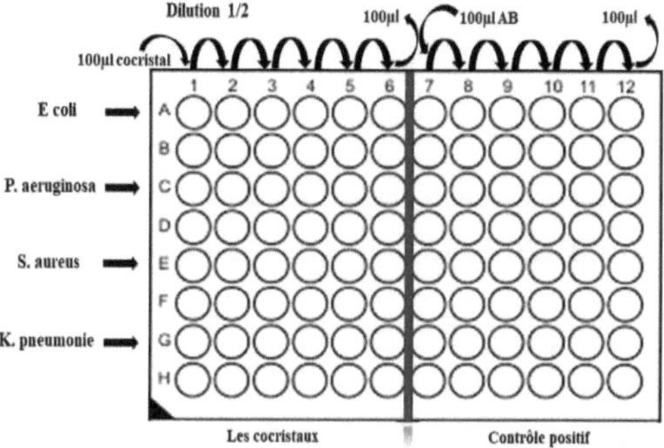

FigureII.10. Preparation of microplates (Original, 2023).

II.6.2.4. Revealing the antibacterial activity of Iodonitrotetrazoluim chloride (INT)

Once incubation was complete, a 0.2 mg/ml solution of Iodonitrotetrazolium chloride (INT) was added to the contents of the 48 wells. Iodonitetrazolium plays the role of an oxidant, acting as an electron acceptor, and is reduced by viable bacteria to produce a coloured product. This yellow reagent turns violet-pink after reduction by the bacteria. A second incubation for 30 minutes is necessary to assess the MIC. After incubation, samples from the unstained wells were seeded and incubated for 18 hours at 37°C to determine the MBC.

CHAPTER III
RESULTS AND DISCUSSIONS

III.1. Introduction

In this chapter, a general discussion of the results obtained will be undertaken. This concerns the synthesis, characterisation and evaluation of the antibacterial activity of the five cocrystals produced on the basis of non-covalent interactions between metronidazole and the five selected coformers.

III.2. Synthesised cocrystals

The cocrystals prepared by wet milling metronidazole with the various coformers chosen according to their chemical structure and non-toxicity, namely L-proline, L-arginine, L-leucine, nicotinamide and paracetamol, underwent a colour change (**figure III.1**) from white to yellow, off-white and brown. This change is visual evidence that suggests a possible phase change during grinding, involving the formation of new compounds, with melting points different from those of the starting compounds. This confirms the establishment of physical interactions between heterosyntons, such as hydrogen bonding, Van der Walls forces and π-π interactions.

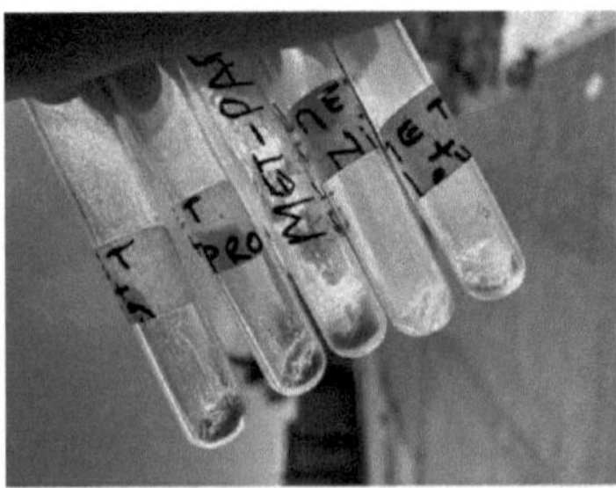

Figure III.1: Physical state and colour of the powders of the various cocrystals synthesised (Original, 2023).

III.3. Characterisation

III.3.1. Differential Scanning Calorimetry (DSC)

The thermograms obtained by differential scanning calorimetry are shown in **Figure III.2**.

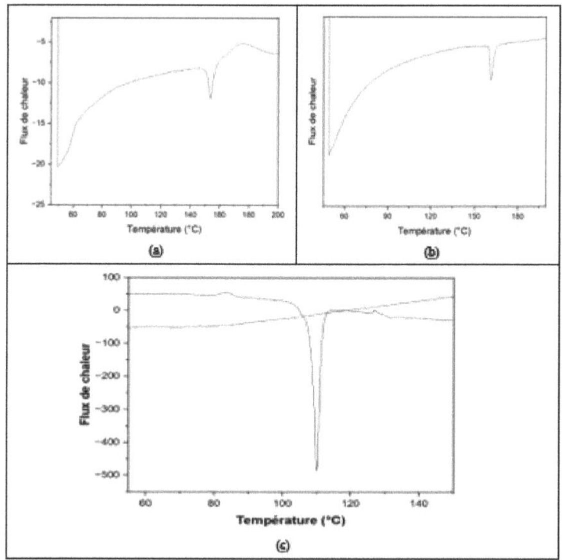

Figure III.2: Thermograms of some of the cocrystals synthesised: **(a)** cocrystal 4, **(b)** cocrystal 5, **(c)** cocrystal 2.

The thermograms show the following:

• MTZ/Pro mixture: formation a co-crystal, resulting in presence of a single accident

It has a thermal melting point intermediate between that of the two pure substances making up the mixture.

• MTZ/Arg mixture: formation a eutectic point, which explained by the presence a single a thermal event with a melting point lower than that of the starting pure bodies.

• MTZ/Nic mixture: formation a eutectic point with a melting point lower than that of the two components pure bodies.

III.3.2. XRD analysis

Figure III.3 shows the diffractograms of metronidazole and the two binary mixtures metronidazole/nicotinamide.

The figure shows that the metronidazole/nicotinamide mixtures produced new peaks at 2θ = 14.38; 18; 25.5; 25.9; 28. This indicates the formation of a new crystalline phase.

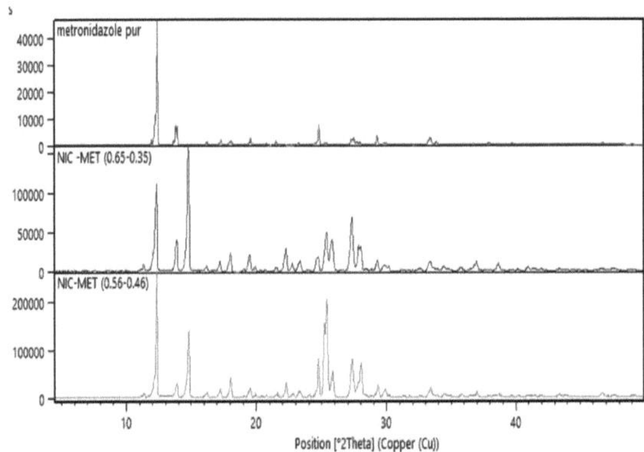

Figure III.3. Diffractograms of metronidazole and cocrystal 2.

III.3.3. Characterisation of synthesised cocrystals by scanning electron microscopy

Figure III.4 shows some SEM micrographs of metronidazole, L-Arginine, L-Proline, cocrystal 4 and cocrystal 5.

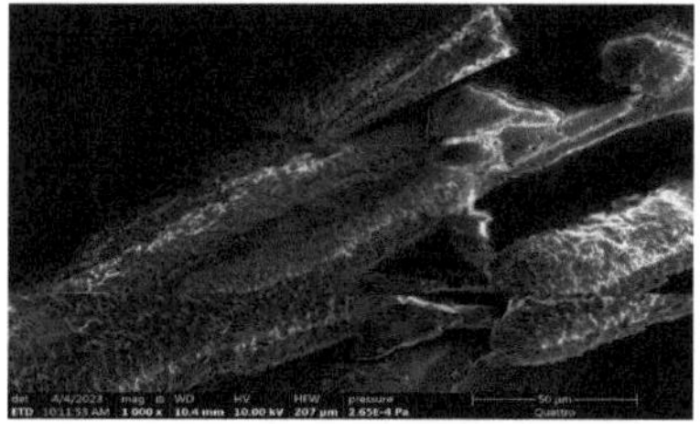

Metronidazole

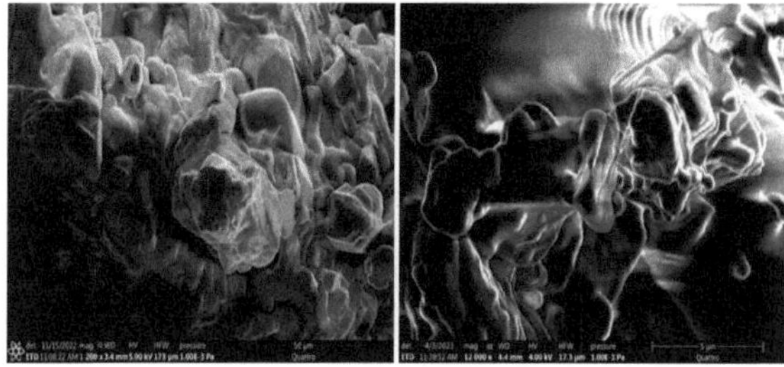

L-Arginine L-Proline

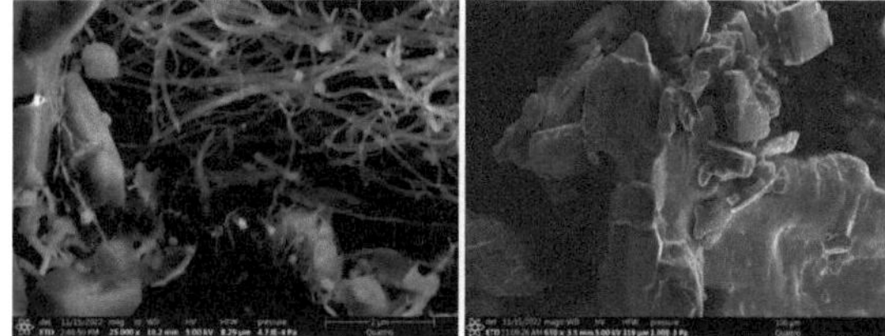

Metronidazole + L- Arginine

Metronidazole + L- Proline Figure III.4: Morphologies of some synthesised cocrystals and their pure bodies.

These micrographs show crystals of different relief and shape. Hexagonal and rectangular, moderately agglomerated platelet structures were observed in pure metronidazole. Co-crystal 4, corresponding to the equimolar binary mixture metronidazole -L- Proline, has a platelet morphology of different shapes and sizes. The morphology of the equimolar metronidazole-L-arginine mixture forming cocrystal 5 is in the form of The new crystalline structures are characterized by the formation of rods of different dimensions in a weak agglomeration. These results confirm those obtained by XRD, which show the formation of new crystalline structures, characterised by a single accident thermal accident.

III.3.4. Characterisation by FTIR

Infrared spectroscopy is an effective technique for intermolecular interactions in solids such as co-crystals and salts. It can be used to determine the nature of the interactions, by means of the different frequencies due to the vibration and elongation of the bonds established by the proton donor and acceptor functional groups.In order analyse some of the cocrystals produced by wet grinding, the samples were finely ground and mixed with potassium bromide. The spectra were recorded with a resolution of 2 cm^{-1}. The infrared spectra of cocrystals 4 and 5 and their pure bodies are **shown in Figure III.5**.

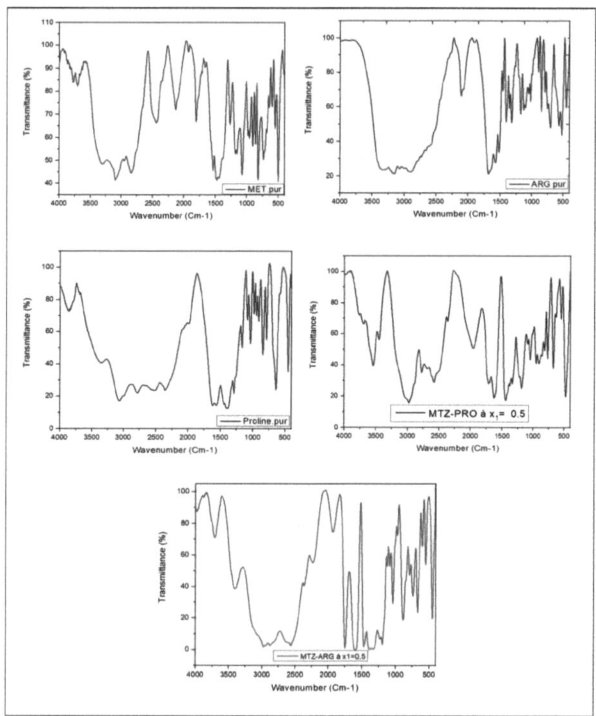

Figure III.5. Infrared spectra of metronidazole, arginine, proline and binary mixtures metronidazole/arginine and metronidazole/proline.

The spectra shown in **(Figure III.5)** show the presence a broad band between 3500 and 3300 cm^{-1}, indicative of association by valence vibration of the intermolecular OH group. Another band was located for the different spectra between 3500 and 3100 cm^{-1}, indicating the valence vibration of the intermolecular NH group.

III.4. Determination of the antibacterial activity of metronidazole co-crystals

III.4.1. Antibacterial activity in solid media

The antibacterial activity of the positive control (metronidazole), of the different PA/coformer combinations at different concentrations and of DMSO as solubilisation solvent used as negative control, was evaluated on four bacterial strains: Escherichia coli, Pseudomonas aeroginosa, Klebsiella pneumoniae and Staphylococcus aureus.

III.4.1.1. Positive control results

Antibiograms in the presence of the positive control, this case pure metronidazole at a concentration of 60 mg/ml, were carried out on the four bacterial strains. The results obtained, shown figure III.6, all reveal the presence of a halo of inhibition of bacterial growth, corresponding to an inhibition diameter of between 6.92 and 13.81. The reactivity sequence of the positive control is as follows:

Klebsiella pneumoniae> Pseudomonas aeroginosa > Escherichia coli> Staphylococcus aureus

Staphylococcus aureus is the microorganism most resistant to pure metronidazole due to the low permeability of the bacterial agent. These results are in line with the literature (N. Islam et al, 2022).

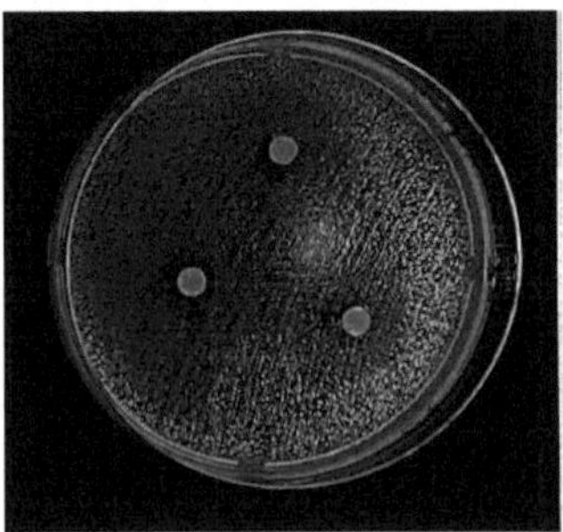

Pseudomonas aeroginosa

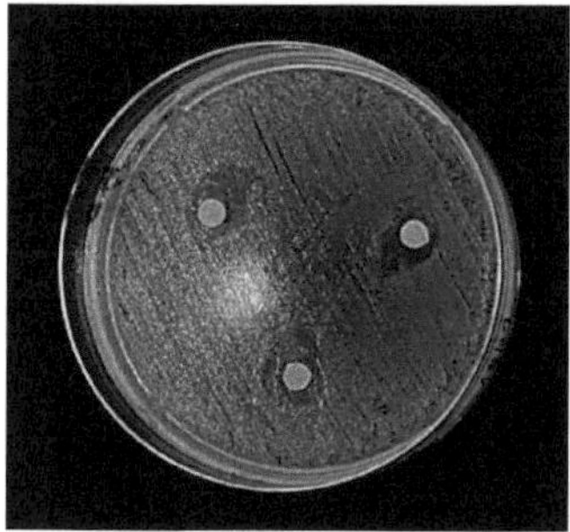

Klebsiella pneumoniae

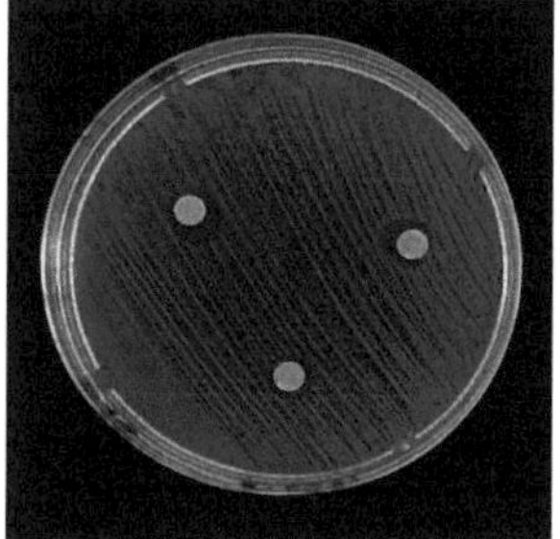

Echerichia coli

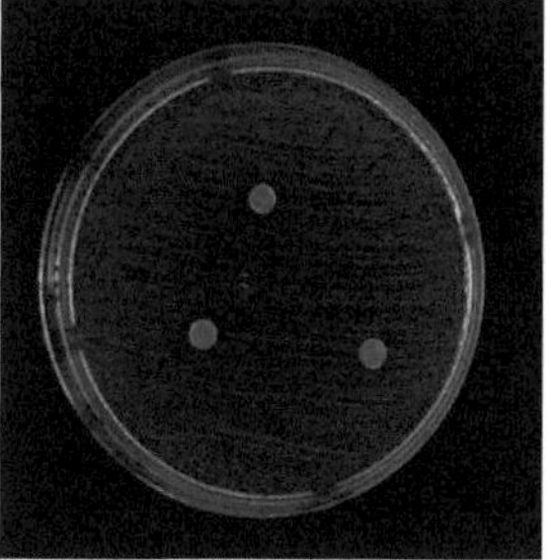

Staphylococcus aureus

Figure III.6. Effect of the positive control on the four bacterial strains tested.

III.4.1.2.Negative control results

Figure III.7 shows that growth inhibition was observed for the negative control. DMSO had no effect on growth of any bacteria on agar. This confirms the correct choice of solvent used for solubilisation.

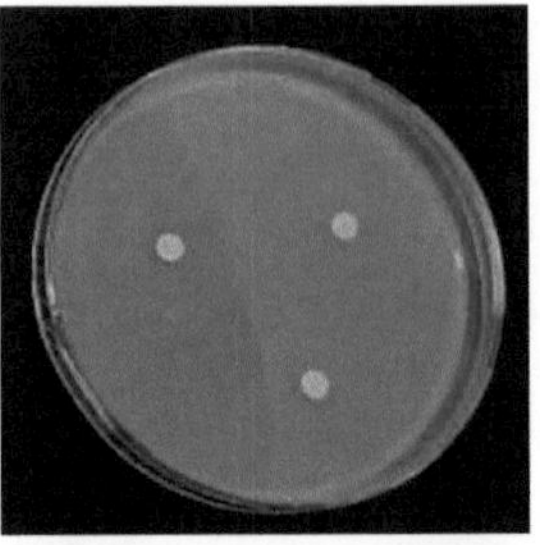

Pseudomonas aeroginosa

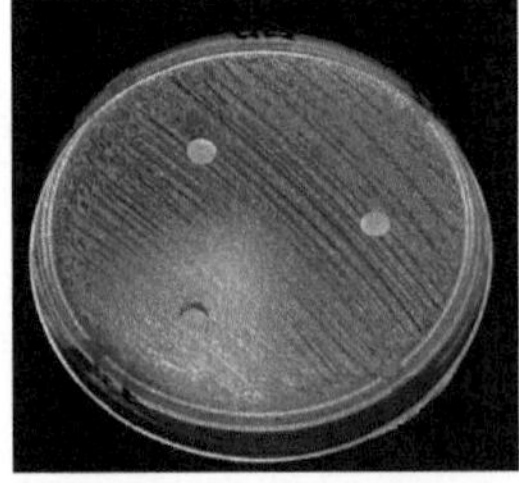

Klebsiella pneumoniae

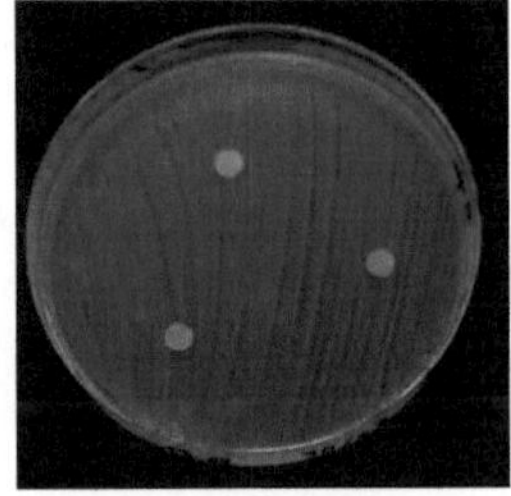

Echerichia coli

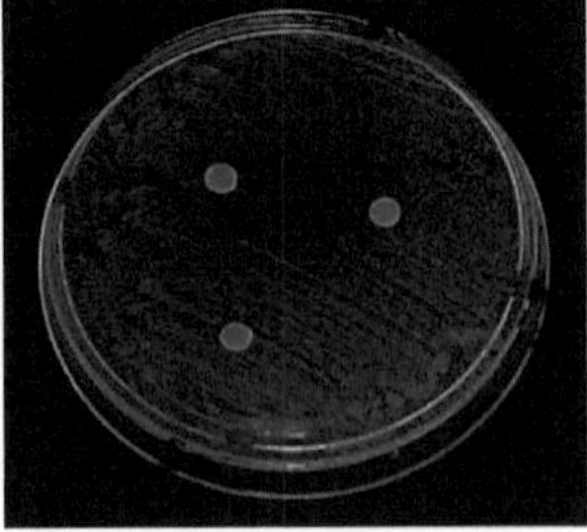

Staphylococcus aureus

Figure III.7. Effect of the negative control on the four bacterial strains tested.

III.4.1.3. Results of tests on antibacterial effect of the five cocrystals synthesised

The antibiograms of the various raw and diluted co-crystals studied formed from PA/coformer physicochemical combination are summarised in (**Table III.1**) and illustrated in (**Figure III.8**).

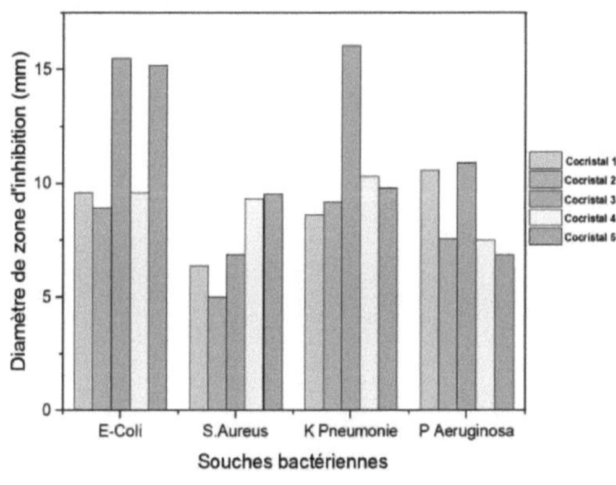

(a)

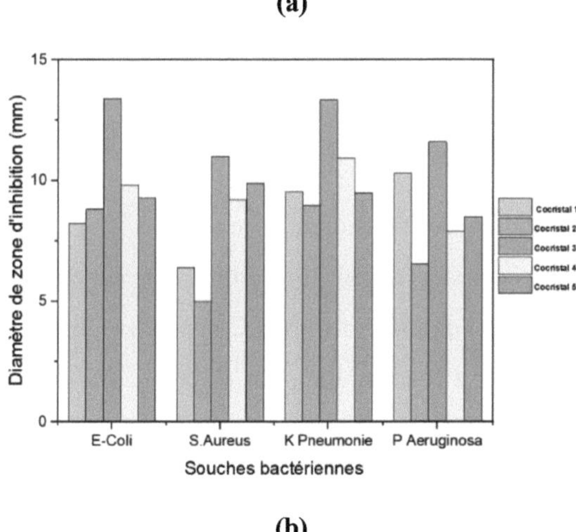

(b)

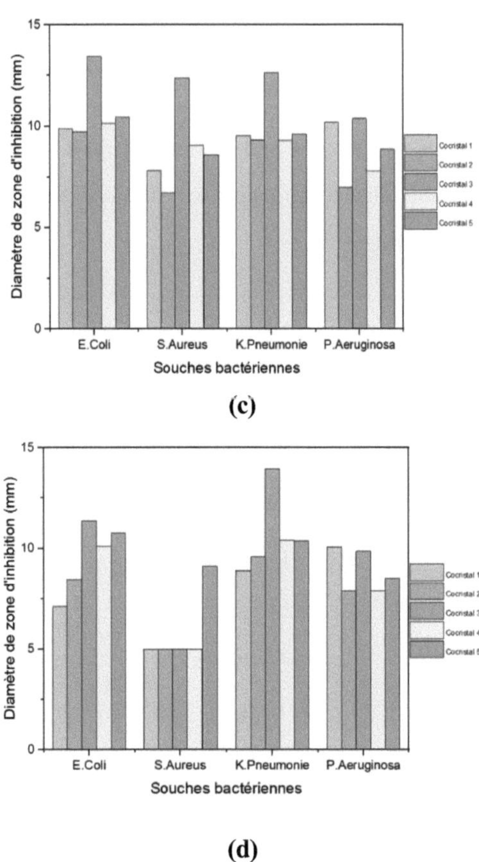

Figure III.8. Histograms of the antibiogram results for the five co-crystals in relation to the bacterial strains tested: **(a)** Comparative histogram of the five raw co-crystals with a concentration of C_0; **(b)** Comparative histogram of the five diluted co-crystals with a concentration of $C_0/2$; **(c)** Comparative histogram of the five diluted co-crystals with a concentration of $C_0/5$; **(d)** Comparative histogram of the five diluted co-crystals with a concentration of $C_0/10$.

The five cocrystals synthesised showed inhibition diameters of between 6 and 16 mm. These values were obtained by antibiogram of gram-negative bacteria. These inhibition haloes can be explained by the chemical structure of these compounds and the nature of the interactions involved in their formation. For example, the highest antibacterial potency was recorded for PA/amino acid combinations, explained by their mechanism action against these microorganisms. Metronidazole and its co-crystals act on DNA, binding to the nucleic proteins of these microorganisms and inhibiting protein synthesis, leading inhibition or death of the microorganism.The histogram obtained from cocrystals of concentration C_0 (**figure III.8.a**) shows a high antibacterial activity of the cocrystal obtained by metronidazole/L- leucine interaction on gram-negative bacteria. It exhibits a halo of inhibition with a diameter ranging from 9.85 to 16.05. The sequence of reactivity of this cocrystal towards the three gram-negative bacterial strains is as follows:

Klebsiella pneumoniae> Escherichia coli > Pseudomonas aeroginosa

Staphylococcus aureus is a strain resistant to the pure antibiotic and its co-crystals. This resistance may be due to modification of the target or enzymatic inactivation.The histogram in Figure III.8.b illustrates the antibacterial activity of cocrystals of concentration $C_0/2$. From this histogram we can see the following:

• Cocrystals with amino acids as coformers showed the highest antibacterial activity in the presence of the four bacterial strains tested. This was reflected in inhibition zone diameters ranging from 7.9 to 13.34 mm.

• The lowest value relates to the antibacterial effect of cocrystal 4 on the bacterial strain P.aeroginosa. The maximum value corresponds to the inhibition caused by cocrystal 3 on the K.pneumoniae.

• Staphylococcus aureus has a sensitivity, corresponding to a halo of inhibition of a diameter of 11 mm, obtained in the presence of cocrystal 3 as bacterial agent. This slight sensitivity can be explained by the structural and electrostatic complementarities between the leucine in this antibacterial agent and the amino acid sequences making up the microorganism's proteins.

The antibacterial effects of the synthesised cocrystals of concentration $C_0/5$ are shown in the histogram in (**figure III.8.c).**

The values obtained for the inhibition zone diameters of the different antibacterial activities of the five antibacterial agents varied between 6.7 and 13.42 mm. Escherichia coli, a gram-negative anaerobic bacterium, is the most sensitive compared with the positive control. This synergistic effect may be due either to the presence of nitrogen and oxygen , proton donors and acceptors capable interacting with the bacterial target, or to the increased permeability of the combined active ingredient, resulting in greater antibacterial activity. These results are in line with the literature (N. Islam et al, 2022).The smallest diameter value of 6.7 mm obtained by the effect of cocrystal 2 on S.aureus can be explained by the chemical structure of this antibacterial agent, which is deficient in proton donor and acceptor groups, reducing the possibility of interaction with the bacterial target.At the concentration of $C_0/10$, shown in the histogram in (**figure III.8.d)**, we can deduce that the bacterial strains are sensitive to moderately sensitive. The values obtained range from 7.1 to 13.93. S.aureus has an antagonistic effect on the positive control, in terms of its resistance to the various co-crystals.

In conclusion

The cocrystals synthesised in this work act in a similar way specific bacterial strains, i.e. they have a synergistic effect towards gram-negative bacteria and an antagonistic effect towards S.aureus, a gram-positive bacterium. The greatest inhibition, resulting in high diameters, is seen in the presence of cocrystal 3 as an antibacterial agent (figure III.9). The antibacterial activities of the four remaining cocrystals are shown in the appendix.

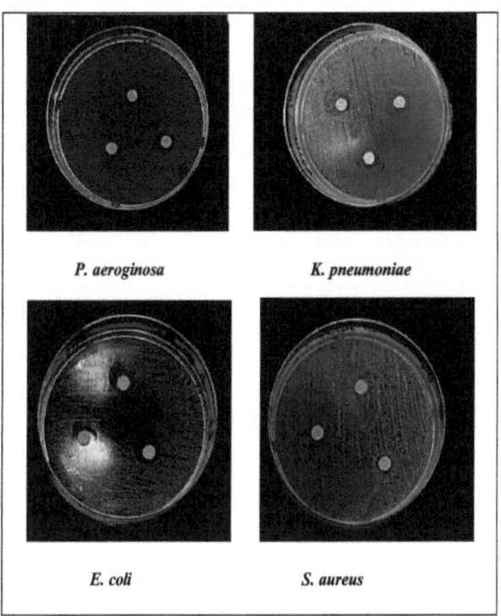

Figure III.9. Effect pure 3-cocrystal on the four bacterial strains tested.

III.5. Determination of minimum inhibitory and bactericidal concentrations (MIC) (CMB)

After determining antibacterial activity on solid media, a complementary study based on the determination of MIC and MBC in the presence of the coloured reagent (INT) proved essential. The MIC is defined as the lowest concentration that inhibits bacterial growth, and is assessed by microdilution on microplates. This revealed some transparent wells and others coloured pink-violet. This coloration is due to an insufficient concentration of the synthesised compound to inhibit bacterial growth, meaning that the bacteria are still alive. The results obtained for MIC and BMC are summarised **in Table III.2** and shown in **Figure III.9**.

Table III.2: MIC (mg/ml) and BMC (mg/ml) obtained by the microdilution method for the four bacterial strains.

Cocrystals	Microbial strains	MIC (mg/ml)	CMB (mg/ml)
	E. coli	3	>3
Cocristal 1	S. aureus	3	>3
	K. pneumonia	3	>3
	P. aeruginosa	3	>3
	E. coli	3.07	>3.07
Cocristal 2	S. aureus	1.535	>1.535
	K. pneumonia	0.38	6.14
	P. aeruginosa	3.07	>3.07
	E. coli	1.545	>1.545
Cocristal 3	S. aureus	3.09	>3.09
	K. pneumonia	1.545	>1.545
	P. aeruginosa	3.09	6.18
	E. coli	1.535	>1.535
Cocristal 4	S. aureus	3.07	>3.07
	K. pneumonia	3.07	>3.07
	P. aeruginosa	3.07	>3.07
	E. coli	3.04	>3.04
Cocristal 5	S. aureus	3.04	>3.04
	K. pneumonia	1.515	>1.515
	P. aeruginosa	3.04	>3.04

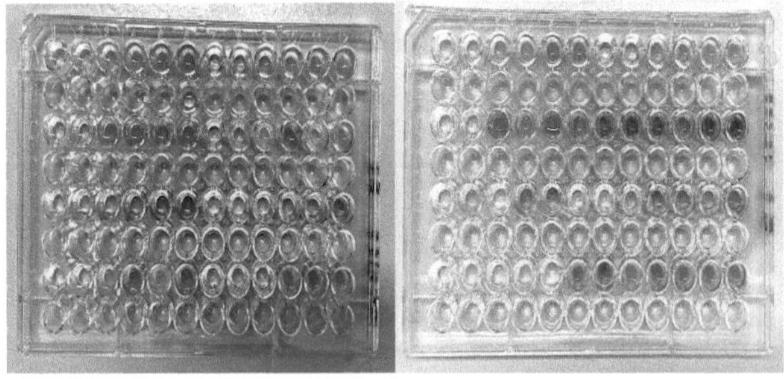

(a) (b)

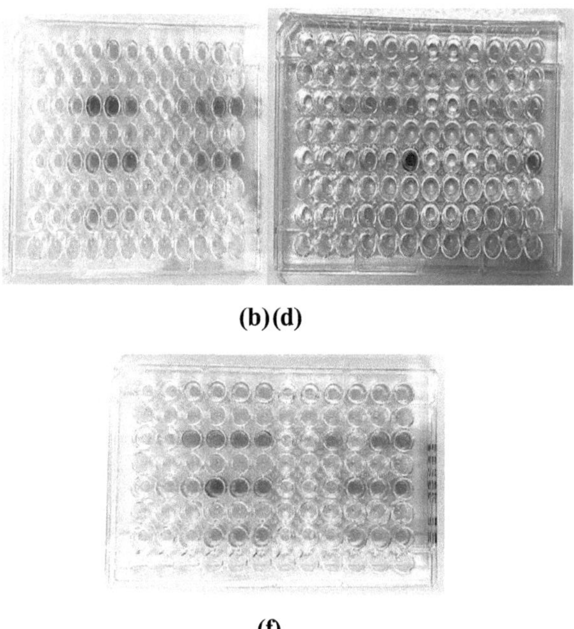

Figure III.10. Determination of MIC and MBC of the five cocrystals studied against the four microbial strains: **(a)** cocrystal 1, **(b)** cocrystal 2, **(c)** cocrystal 3, **(d)** cocrystal 4, **(e)** cocrystal 5.

According to the literature (N. Djabou et al, 2013) the bacteria are :

- Resistant when their MIC is between 25 and 50 mg/ml;
- Moderately sensitive (+) for a value between 3 and 12.5 mg/ml ;
- Sensitive (++) for a MIC between 0.4 and 2 mg/ml ;
- Extremely sensitive when their MIC is less than or equal to 0.2 mg/ml.

The minimum inhibition concentrations evaluated in this study ranged from 0.38 to 3.09 mg/ml.The bacterial strains tested were sensitive to moderately sensitive. The sensitivity sequences of the bacteria are as follows:

- Antibacterial agent metronidazole/paracetamol: all the bacterial strains tested showed the same sensitivity to this agent, with a MIC of around 3mg/ml. In comparison with pure metronidazole, this agent had an indifferent effect on E.coli and P.aeroginosa and an antagonistic effect on Staphylococcus aureus and K.pneumoniae.

- Antibacterial agent metronidazole/Nicotinamide: K.pneumoniae is the most sensitive, with an MIC of around 0.38 mg/ml. It is an anaerobic bacterium sensitive to imidazole antibiotics and their derivatives. It also has a synergistic effect on P.aeroginosa, S.aureus and K.pneumoniae.
- Antibacterial agent metronidazole/leucine: gram-negative bacteria are sensitive to this agent, with an MIC of 1.545 mg/ml. S. aureus, a gram-positive bacterium, did not show the same sensitivity. These results are in agreement with those obtained using the disk diffusion method.
- Antibacterial agent metronidazole/ Proline: E.coli is the most sensitive with an MIC of 1.54 mg/ml. This synergistic effect explains why the antibacterial effect of this agent is better than that of pure metronidazole. On the other hand, it has an antagonistic effect on staph, a bacterium that is generally resistant to imidazole antibiotics.
- Antibacterial agent metronidazole/arginine: In comparison with pure metronidazole, this agent has an indifferent effect on E. coli and P. aeroginosa and an antagonistic effect on S. aureus and K. pneumoniae.

The MBC allows us to deduce the nature of the antibiotic: it is bactericidal if this concentration is equal to the MIC or the MIC/CMB ratio is less than or equal to 4. It is bacteriostatic when its MBC is greater than its MIC and the MIC/CMB ratio> 4. We can deduce from our results that at the concentrations taken, our antibacterial agents behave as bacteriostatic, with the exception of cocrystals 2 and 3, which are bactericidal at a concentration of 6 mg/ml.

CONCLUSION

This work is part of a PRFU research project looking at the co-crystallisation of active ingredients to improve their physico-chemical properties and increase their biological activity.The active ingredient chosen for this study is metronidazole, an antibiotic and antiparasitic of the 5-nitroimidazole family, active on anaerobic bacterial strains. This antibiotic was co-crystallised by combining it with various organic molecules that have no therapeutic effect and act as coformers. These include paracetamol, nicotinamide, L-Leucine, L-Proline and L-Arginine. A number of cocrystals synthesised by wet milling of the active-coforming principle in a 1:1 stoichiometric ratio were characterised by DSC, PXRD, SEM and FTIR. The antibacterial activity of the cocrystals produced was determined by two complementary methods: the solid-state diffusion method and microdilution in microplates, enabling the diameter of the zones of inhibition and the minimum inhibitory and bactericidal concentrations to be assessed, respectively. The bacterial strains chosen were Escherichia coli, Pseudomonas aeruginosa, Klebsiella pneumoniae and Staphylococcus aureus.The results obtained from differential scanning calorimetry show the existence of a single thermal accident, explained by the formation of a compound (cocrystal or eutectic point) by non-covalent physical interactions. These results were confirmed by PXRD with the appearance of new peaks, by SEM with the change in morphology and by FTIR with the appearance of an intense band at around 3500 cm^{-1}, corresponding to intermolecular hydrogen bonding.The results of the antibacterial activity obtained using the two methods used are comparative. The antibacterial activity of the cocrystals was assessed in vitro using the disk diffusion method, for the four bacterial strains tested, follows the reactivity sequence:

Klebsiella pneumoniae> Escherichia coli > Pseudomonas aeruginosa> Staphylococcus aureus

From these results, we can deduce that the cocrystals synthesised in this work act in a similar way towards specific bacterial strains, i.e. they show a synergistic effect towards gram-negative bacteria and an antagonistic effect towards Staphylococcus aureus, a gram-positive bacterium. Cocrystal 3 is the most powerful bacterial agent, with inhibition diameters between 10 and 16 mm.

Determination of the MIC allowed us to deduce that the co-crystals from the metronidazole/amino acid and metronidazole/nicotinamide combinations acted synergistically with the pure active principle on most of the bacterial strains tested. This result can be explained by the nature of the functional groups present, which are structurally and electrostatically complementary to the active sites in the bacteria's DNA. These functional groups target bacterial proteins, inhibiting their growth.The lowest value of the minimum inhibition concentration of 0.38 mg/ml was recorded in the presence of the antibacterial agent metronidazole/ nicotinamide against Klebsiella pneumoniae. Klebsiella pneumoniae is an anaerobic bacterium sensitive to imidazole antibiotics and their derivatives.The BMC results obtained show that at the concentrations taken, our antibacterial agents behave as bacteriostatic, with the exception of cocrystals 2 and 3, which are bactericidal at a concentration of 6 mg/ml.

Looking ahead

Further studies could be envisaged to explore in greater depth some of the points raised in this manuscript.

It would therefore be interesting to :

- Synthesising co-crystals from a variety of antibiotics, belonging to different families and acting on different bacterial targets, in the presence of coformers with different chemical structures, to combat the spread of resistant bacterial strains that are constantly on the increase.

- Evaluate their solubility in a range of alternative solvents, such as ionic liquids.
- Determine their antibacterial activity against multi-resistant bacteria.

REFERENCES

A

Aitipamula S, Banerjee R, Bansal AK, Biradha K, Cheney ML, Choudhury AR Polymorphs, salts, and cocrystals: what's in a name? 2012;12(5):2147-52.

B

Banerjee R, Bhatt PM, Ravindra NV, Desiraju G, design. Saccharin salts of active pharmaceutical ingredients, their crystal structures, and increased water solubilities. 2005;5(6):2299-309.

Bashimam M, El-Zein H. Pharmaceutical cocrystal of antibiotic drugs: A comprehensive review. 2022:e11872. Bauer J, Spanton S, Henry R, Quick J, Dziki W, Porter W. Ritonavir: an extraordinary example of conformational polymorphism. 2001;18:859-66.

Berry DJ, Steed J. Pharmaceutical cocrystals, salts and multicomponent systems; intermolecular interactions and property based design. 2017;117:3-24.

Billot P, Hosek P, Perrin M-A, Development. Efficient purification of an active pharmaceutical ingredient via cocrystallization: From thermodynamics to scale-up. 2013;17(3):505-11.

Biscaia IFB, Gomes SN, Bernardi LS, Oliveira P. Obtaining cocrystals by reaction crystallization method: Pharmaceutical applications. 2021;13(6):898.

Bouskraoui M, Zouhair S, Soora N, Benaouda A, Zerouali K, Mahmoud M. A practical guide to pathogenic bacteria. 2017.

Bučar D, Filip S, Arhangelskis M, Lloyd G, Jones. Advantages of mechanochemical cocrystallisation in the solid-state chemistry of pigments:

colour-tuned fluorescein cocrystals. 2013;15(32):6289-91.

C

Ceruelos AH, Romero-Quezada L, Ledezma JR, Contreras L. Therapeutic uses of metronidazole and its side effects: an update. 2019;23(1):397-401.

D

Dangoumau J. General Pharmacology. Edition2006. 2006.

Djabou N, Lorenzi V, Guinoiseau E, Andreani S, Giuliani M-C, Desjobert J-M. Phytochemical composition of Corsican Teucrium essential oils and antibacterial activity against foodborne or toxi-infectious pathogens. 2013;30(1):354-63.

H

Hachem CY, Clarridge JE, Reddy R, Flamm R, Evans DG, Tanaka SK. Antimicrobial susceptibility testing of Helicobacter pylori comparison of E-test, broth microdilution, and disk diffusion for ampicillin, clarithromycin, and metronidazole. 1996;24(1):37-41.

I

Islam NU, Umar MN, Khan E, Al-Joufi FA, Abed SN, Said M, et al. Levofloxacin cocrystal/ salt with phthalimide and Caffeic acid as promising solid-state approach to improve antimicrobial efficiency. 2022;11(6):797.

K

Kaakoush NO, Asencio C, Mégraud F, Mendz G, chemotherapy. A redox basis for metronidazole resistance in Helicobacter pylori. 2009;53(5):1884-91.

Kablan B, Adiko M, Abrogoua D. In vitro evaluation of the antimicrobial activity of Kalanchoe crenata and Manotes longiflora used in ophthalmia in Côte d'Ivoire. 2008;6(5):282-8.

Korotkova EI, Kratochvíl B. Pharmaceutical cocrystals. 2014;10:473-6. Kumar N, Rohilla RK, Roy N, Rawat D, letters mc. Synthesis and antibacterial activity evaluation of metronidazole- triazole conjugates. 2009;19(5):1396-8.

L

LA MJ-, Bahi C, Dje K, Loukou Y, Guede-Guina F. Study of the antibacterial activity Morinda morindoides (Baker) milne-redheat (rubiaceae) acetatique extract (ACE) on in-vitro growth of Escherichia coli strains Study of the antibacterial activity of Morinda morindoides (Baker) milne-redheat (rubiaceae) acetatique extract (ACE) on in-vitro growth of Escherichia coli strains. 2008.

Li J, Hao X, Wang C, Liu H, Liu L, He X, et al. Improving the solubility, dissolution, and bioavailability of metronidazole via cocrystallization with ethyl gallate. 2021;13(4):546.

N

Neurohr Cm. Elaboration of pharmaceutical cocrystals by CO2-assisted processes: Bordeaux; 2015.

R

Rediguieri CF, Porta V, Nunes DS, Nunes TM, Junginger HE, Kopp S. Biowaiver monographs for immediate release solid oral dosage forms: metronidazole. 2011;100(5):1618- 27.

S

Singh M, Barua H, Jyothi VGS, Dhondale MR, Nambiar AG, Agrawal AK. Cocrystals by Design: A Rational Coformer Selection Approach for Tackling the API Problems. 2023;15(4):1161.

T

Thayyil AR, Juturu T, Nayak S, Kamath S. Pharmaceutical co-crystallization: Regulatory aspects, design, characterization, and applications. 2020;10(2):203.

W

Whelan J, Hale J. Bactericidal activity of metronidazole against Bacteroides fragilis. 1973;26(6):393-5.

Y

Yaghoubi S, Zekiy AO, Krutova M, Gholami M, Kouhsari E, Sholeh M, et al. Tigecycline antibacterial activity, clinical effectiveness, and mechanisms and epidemiology of resistance: narrative review. 2021:1-20.

Z

Zheng K, Gao S, Chen M, Li A, Wu W, Qian S. Color tuning of an active pharmaceutical ingredient through cocrystallization: a case study of a metronidazole-pyrogallol cocrystal.2020;22(8):1404-13.

Zheng K, Li A, Wu W, Qian S, Liu B, Pang Q. Preparation, characterization, in vitro and in vivo evaluation of metronidazole-gallic acid cocry.

TABLE OF CONTENTS

Printed by Books on Demand GmbH, Norderstedt / Germany